International Cataloging Data in the Publication (ICDP)

C972.d   Cunico, Marlon Wesley Machado,

How to use advanced finishing techniques
to increase your value and multiply your profit by 20 /
Marlon Wesley Machado Cunico; Concep3D Pesquisas
Científicas Ltda; Curitiba, 2019,

ISBN: 9798628249949

1.3D   Printer;   2.   finishing   3.   Technology.
Título

CDD 620

Systematic Cataloging Index

1. Engineering and applications 620

1st Edition - 2020

Printed in Brazil

Concep3D Pesquisas Científicas Ltda

298 Pedro Ivo street ap 23

80010-020 Curitiba, Brasil

http://www.concep3d.com

How to use advanced

Finishing techniques to

INCREASE YOUR VALUE

AND

MULTIPLY YOUR

PROFIT BY 20

Responsibility Term

All the strategies and information that you will read in this book, and what I learned when I started using advanced finishing techniques to increase the value of my products are fruit of my professional experiences in the area, in addition to more than 15 years of scientific research in Additive and advanced manufacturing.

Although I did my best to ensure the accuracy and the highest quality of this information, in way that all the techniques and methods taught here are highly effective for anyone who is not inclined to learn and put the suitable effort required to apply them as instructed, these methods and information is not possible to be learned theoretically, but only in the practice.

The strategies and information presented here are for all, but not for anyone. You need to be willing. In addition, your particular situation may not be adequate perfectly to the methods and techniques taught in this guide. So, you can use it by adjusting the information according to your specific need and, for that reason, results may vary from person to person. There is no guarantee, there is only one experience and testimony of thousands of customers successful thanks to this method.

My name is Marlon Cunico, I born in Londrina (Brazil) in 1984. My parents moved to Curitiba (Parana Capitol) looking for a better life.

Both of my parents came from humble family, where only my mother had come to university. As a consequence, life was hard and unfair several times .

My father is a wise man who graduated in Mathematics and Physics, even though he had to work in 2 or 3 different jobs to support our family. In addition, my mother used to work 60 hours per week in order to pay my studies.

I studied as hard as it was possible because we could not afford to pay private University. Therefore, I was accepted in one of the most prestigious Engineering University in Brazil on year before I finish high school.

After 7 years , I finished Engineering, Mastering in Engineering and got my Ph.D. in Robotics and Advanced Manufacturing.

In 2011, I was working in the department of innovation and engineering of one of the biggest automotive Industries in the globe.

Years before, I had started working with Rapid prototyping technologies (which become 3D printing after some time). I was so fascinated about such amazing technologies that **I invented and patented a brand new process by the end of my Masters**. I had realized that these technologies would change the world as we had known. They changed indeed, and 9 years after that moment, several researchers and experts indicate that 3D printers boost the 4th industrial revolution.

Therefore, I took one of the most difficult decisions of my life. I founded my first company. The main goal of that company was to produce prototypes and low scale production batches applying 3D printing technologies.

Amazing, isn't it? That is right.

But in 2013, the biggest popularization of 3D printing technologies happened and thousand of 3D printing enthusiasts have got exactly the same conclusion that I had before.

We started playing the game using the tools that big companies (majorly manufactures of 3D printing) sold to us. We optimized 3D printing parameters, used and developed closed chambers and achieved 3D printing results that were unbelievable.

But there was just one problem ……..

Things looked like a roller-coaster. Part of clients that understood foundations of 3D printing were ok with our deliveries. But major of clients who did not give a damn about 3D printing techniques and their parameters were extremely unhappy with results that in general way were outstanding.

I saw around 82% of companies from that time closing their doors and then I decided to dedicate my research career to advanced manufacturing.

I realized that everyone was playing with wrong rules which were defined by big corporations that just wanted to sell you their machines. The tools that were provided were obsolete and aim to validate the advantages of their equipments against the competitors.

Then, I start developing a guideline which maps customers need and ensures suitable results for my clients. This was the moment when the **table start turning over** .

Turning the table

I was professor and researcher at University in addition to managing company. Then **I launched new research** which aimed to investigate **productivity and finishing techniques** of 3D printing technologies and classic technologies altogether.

We identify that 3D printing technologies could be more productive than classic technologies. But **severe changes were needed** to be implemented in order to achieve such an amazing result.

We started to collect, modify and develop different techniques upon Lean manufacturing, Toyota techniques,

Six Sigma, Open source technologies, Jewelry, foundry and casting among others.

As a result, my company survived to the storm, increased profits in 50 times and obtained more than $500k in funds and projects per year.

After solving the most dangerous threat of our company. I have been seeing the same problem happen over and over again, killing thousands of companies nowadays. Therefore **I decided to share and teach other people and entrepreneurs** about what I worked for me.

In several events people were learned how to change their mind and improve their services to another level.

Now, my goal is to bring this message to as many people as possible. And for that reason I am sharing the essence of the best practices in this book.

BE BOLD, MAKE IT HAPPEN !!!!!!!!!!

→ A step further

Story is a complicated thing. I made what it was possible to summarize mine in here. But if you want to know more about how to make things stunning, access:

www.makeitstunning.com

Your learning and self development is directly proportional to the energy, focus and immersion in the subjects you want to learn. It is not just a valid observation for advanced manufacturing and finishing techniques, it is also true for everything that you intend to learn.

So, one of the things you will need to do is to create your "own world", where in much of the time you will practically breathe Product Development, advanced manufacturing and 3D printing technologies.

This book will be your central guide, but read a good part of it and spend the rest of your day watching TV is not going to help quite much. You'll need more control about the inputs information you are handling, especially when you finish this book and start applying strategies in your business.

So I would like to share with you some tips to facilitate your immersion process in that Entrepreneurship universe:

1. A step further

Take time to become an overachiever! Watch the complementary videos I prepared for you expand your knowledge. In them, I count the my story and what brought me here, comment excerpts from the book, I give examples and much more!

2. Take notes

Try to read calmly and summarize the main contents in a notebook. Do it in such a way that you can explain the content learned to another person.

3. Subscribe to my YouTube channel

https://www.youtube.com/channel/UC8QFph7-R2YERpN5cEAl1Mw

There, you will find great views of the world of advanced manufacturing.

4. Follow my Instagram profile

https://www.instagram.com/concep3D

Daily pills of wisdom and backstage of everyday life an entrepreneur.

5. Subscribe to my newsletter

For you who are a fan of audio content, look for Marlon Cunico in his favorite podcast application.

6. Create study group

Refer this book to others and form a group of studies. Discuss the book's content with friends, colleagues and close people helps you to go beyond. Remember: if you want to go fast, go alone, otherwise if you want to go far, go accompanied.

I believe that this book is going to be very useful for quite many people. On the one hand, it is extraordinary for those who have already work with 3D printing services or have their own 3D digital businesses. On the other hand, this book is terrific for the ones who neither work with 3D printing nor have 3D digital businesses.

In addition, the strategies that I expose here go further and helps to create an infinity of new sort of businesses. Therefore, I have several case studies in different segments, such as health care, medical application, industry, games, dentistry, product design, maintenance, services, robotics, engineering, architecture and many others.

I will be very glad if you finish this book and realize the multiple opportunities and potential application of this methodology.

In the last part of this book, I will also presents a product strategy which I used in my company rising. This strategy is so unbelievable that it allowed me to:

- Have the highest profit margin which I have ever seen in the planet
- Have a lean structure, with few employees (where I was not even physically present most of the time)
- Kickoff with an investment which is so small that everyone could implement
- Make your clients extremely happy and faithful
- Have outstanding results

You will discover this strategy in the 3rd part of this book. But first, I will have to teach you 2 very important things in order to understand why this strategy is so worthy.

1. What is the sort of quality (product or service) that your customer see

You will learn to use, in your business or personal designs, a manufacturing strategy that will bring **measurable and quantitative results**.

Measuring each action and project that you deliver will allow you to **scale your investment in a super reduced risk environment** . Therefore, it will help your business to improve in addition to create new branches and new opportunities.

The best part is that **you don't need to begin with a millionaire budget**. Invest wisely a very small budget is enough to improve your projects and increase results.

One thing I can ensure to you, after finishing this part of book, you are never going to see the 3D printing services from the same point of view. For 99% of 3D printing user (beginners to experts), advance manufacturing is a big black box where you send a STL to one side and receive a 3D printing object from the other side.

After finishing this part of book, you will be part of 1% of people who understand what happens inside the mysterious black box.

After explaining that to you, I will teach you:

In this part of the book, I will explain to you **how to create a quality driven strategy** in a step-by-step guidance. This strategy unbelievably works for those who either have experience or are beginning from zero point.

Do you want to know why this strategy is so extraordinary? Because it allows you to:

- Optimize your job time
- Deliver an accurate, precise and high quality product
- Improve in characteristics which actually are relevant for your clients
- Increase your value in accordance with your client satisfaction
- Create metrics to always reach the peak performance from the customer point of view
- Increase your profits – a lot

Well, do you know why I organized this book in this fashion way? Because although you haven't got your own 3D printing office yet, the best thing you can do now is to learning fabrication strategies that actually work.

Imagine one thing, If you don't know how to see customer needs, fabricate outstanding objects or design high valuable products; Just giving you the magic strategy on a silver plate will make absolutely no sense. Because you won't be able to move from wherever you are.

On the other hand, what I show in this book is so powerful that, I am sure, it will be almost impossible to ignore this new projects and design, even if you already have yours.

I have countless case studies of people who created new business units or pivoted your business to this new strategy, after find out what I'm going to teach you.

I truly believe that the content of this book can transform your way to design and fabricate products in 3D printing. I say so because this content transform my company and thousands of peoples whom I mentored during the years.

Therefore, **it is very important you consume the contents of this book in the sequence in which it is presented to you**.

Believe in me, I have spend quite many hours in order to find the best way to present the worldwide best practices which is presented to you so you can **consume, learn and apply as efficiently as possible**.

And, moreover: **read, learn, apply and tell people around you!**

1

What is the sort of quality (product or service) that your customer see

How are extraordinary results possible to be reached?

The results obtained by people who learned the best practices that you will learn in this book are extraordinary! Best of all, I honestly believe that anyone that is really committed to learn and apply the strategies proposed in this book will reach good rewards.

I do not identify it by intuition, but according to over 2000 case studies which I cataloged from scientists, engineers, hobbyists, businesses which partially implemented tools and strategies that I teach you in this book.

"Insanity: doing the same thing over and over again and expecting different results."

Albert Einstein

As a consequence, many of them reported increase of value over 20 times, being an extraordinary result.

Nonetheless, don't mess results up. Results like this in the first campaign is abnormal. The proposed strategy tend to be more like a transformation of company philosophy. In this case, let's call it **Customer Centric Transformation**.

What do I mean by that?

I mean that not everyone will reach this sort of result.

Do you know why?

Because although it is possible, doing what you have to do to achieve such results is not easy. In this book I'm going to give the map and point the direction

Nevertheless, the effort and dedication to study, learn and apply the tools depends entirely on you.

 As the old proverb says : "**You can lead a horse to water, but you can't make it drink** ".

And unfortunately, not everyone has the willing to do their part, to enter the field and not to give up on first stumbling block or obstacle.

Reading this book and continuing to be seated in the chair, doing nothing, will not make the results appear magically for you.

 But if you have enough will to both roll up your sleeves and go to the combat, this book will give you the path that many have already taken to get there.

So, now that all the cards are on the table, how is it possible to have extraordinary results?

Not all customers are the alike

First of all, you will have to understand that there are 3 type of customers that you will service or convert. The comprehension of these customers will allow you to create suitable strategies to better satisfy their essential hidden need. Therefore, it is possible to reach peak of efficiency for most or all clients in accordance with your strategy flexibility.

Types of Customers

Let's imagine that all the customers are like an Iceberg. Most of times, only the tip of iceberg is visible (out of water) . In this case, the iceberg tip illustrate our first type of customer. Nonetheless, this customers is the most wanted among the competition , which intends to convert it and build customer loyalty.

Well, you must have been imaging which type of customer is this?

"People don't know what they want until you show it to them. That's why I never rely on market research."

Steve Jobs

The 3DP EC have 3 main benefits:

- They know exactly what they want about 3D printing services
- They know what to expect in terms of roughness, finishing, dimension distortion and strength
- They normally know 3D printing parameters and prepare STL and supports in the direction it is expected to be printed

For this customer, you have to put no efforts in convincing about the wonderful benefits of 3D printing. Normally, you don't even have to look for they , because they are always looking for new technologies and new advances in 3D printing field.

As a consequence, fabrication strategy and post processing are minimized, as they already know the application in which they will apply the 3D printing object. The roughness and strength are well known and you don't have to put any effort in discovering how to satisfy their needs.

That is exactly why most of 3D printing services kill themselves for the tip of the iceberg. Therefore, there is where your competitors are working hard to steal your clients. Therefore, it will lead you to compete in price, reducing your profit margin and jeopardizing your relationship with all the other potential customers of yours.

Thus, 3DP EC tend to have 3 major disadvantages :

- High competition
- Lack of loyalty
- Small range of market and profit margins

In other words, they lead your business to a red ocean. Therefore, if your service segment is still small , lacking of competitors, you will achieve very interesting results by attending this sort of customer.

On the other hand, segments where the competition is high force you to differentiate or quit. Both of the cases are likely to happen and it has already taken around 80% of companies out of the game since 2013.

Therefore, I will teach you how to dive deeper to reach the second type of customer.

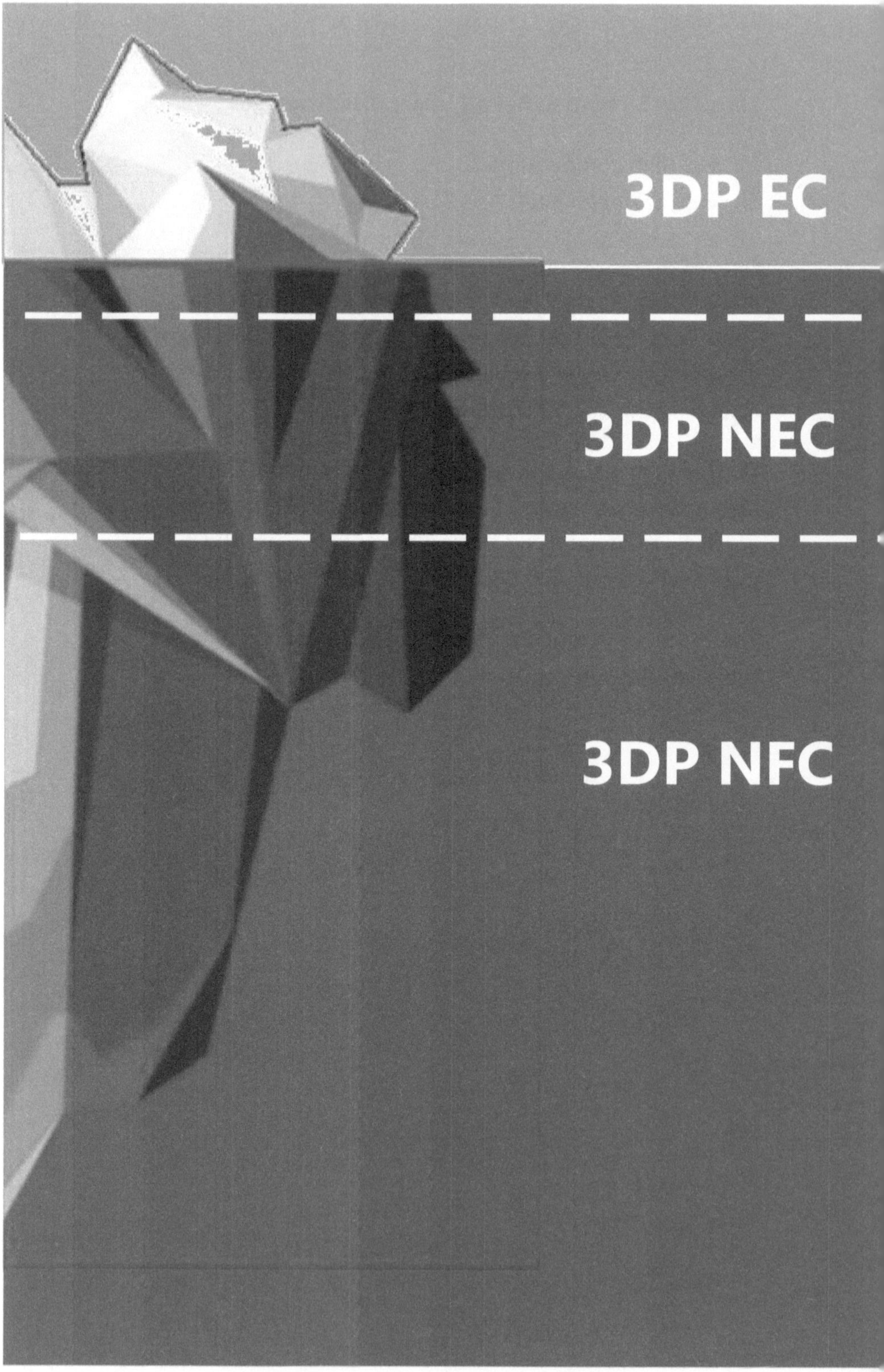
3DP EC
3DP NEC
3DP NFC

What is the main characteristic of 3DP NEC?

- They believe and have conviction that need 3D printing services. However, they do not know neither how 3D printing works nor what objects are expected to be delivered .
- They are motivated by 3D printing and know about their product designs

This type of customer is often a 3D printing enthusiasts which believe that 3D printing objects be "Perfect". In general , they are professionals that use 3D printing technologies for development acceleration, design validation, prototypes and mockups, architecture models and for marketing and advertising. In addition, medical doctors and dentists are also included in this type of customer .

It is possible to see that for each 3DP EC, we can find 20 3DP NEC. And it makes the second piece of the iceberg 20 times bigger than the tip of iceberg.

Thus, the magical secret is to find the strategy that will reach these customers. Obviously, the tools which are used for the first type are completely different from the tool which are used for the second type of customers.

Let's make an illustration, to find rocks on the tip of iceberg, no diving equipment is required. In this case, you do not even need to know how to swim.

Now on the second piece of iceberg, some techniques are required. You will have to know how to swim and sometimes use diving mask to reach submerse rocks.

Ok. What techniques I can use to deliver high quality products for this sort of customer? I will show you those tools later in this book. Moreover, I tell you in advance that this techniques are simple, but difficult.

How is that possible? You must have been asking yourself.

I will explain to you. Imagine that you want to train to run a marathon. That is very simple because there are tons of apps which can help to know what is needed. Nonetheless, it is not easy, isn't it? It is very hard because you will have to spend quite much energy, dedication and time on the training in order to be prepare to run the marathon.

Let take a look in the advantages of the 3DP NEC:

- They are found in large scale
- There are less competitors fighting for them
- Therefore, price start not being the deal breaker and you can increase your profit margins in order to invest in your business (labor, equipments, technologies, et cetera)

On the other hand, what are the disadvantages of 3DP NEC ?

- You will need high profile technique and energy to identify , materialize and deliver what they want

Alright, here you can see that there are disturbances and barriers that disturb you in the process. So, if you want to dive deeper and explore the larger part of this iceberg, thing will become even harder.

In the case you aim to reach an unbelievable larger group of customers, let's go deep and understand the third type of customers.

What attracts a customer who is not familiar with 3D printers?

- He need to fabricate and validate their design, but they have no idea about how to do so
- They are not looking for 3D printing services

There are people who need to take care of their health even though they do not know what is needed to do so. Likewise, there are people who need 3D printing, advanced manufacturing and advanced finishing to validate, launch and sell their products, but they do not know what they need yet.

In this case, each customer from the tip of iceberg (the first part which every competitor kills to have) correspond to 50 3DP NFC.

What are the benefits of 3DP NFC?

- They are found in even larger scale (50 -3DP NFC per each 3DP EC)

- This segment has even less competitors
- Therefore, as your business will be driven by the value which the customer sees, you can increase you profit margins up to a fair cost

As the 1st and 2nd type of customers, 3DP NFC have their disadvantages too:

- You will need to apply even more energy, patience and high profile techniques in order to reach them

Well, talking by experience of who discover an strategy that actually works, I always aim at the 3DP NEC and 3DP NFC. Apart from the moments that there are no competitors, I almost never dispute the 3DP EC because these moments are rare.

It is amazing to talk with either 3DP NEC or 3DP NFC because , it this zone, I have no competitors . In addition, I can offer disruptive solutions, making the customers extremely happy, offering suitable techniques and obtaining huge profit margins.

Now that you know how to reach this two type of customers (3DP NEC and 3DP NFC),you will need to understand the 2 models of advanced 3D printing services.

3

The 2 models of advanced 3D printing services

Every business, no matter what stage it is in (either the one who are taking baby steps or the giant of the market), wants and need to have more customers.

But, to achieve this, what is usually done? People invest in equipment, marketing, labor and infrastructure. In other words, it is common sense that it is necessary to improve quality and productivity to bring more customers and create loyalty relationship. And, of course, you want that, don't you?

Technical branding Services

When the matter is about advanced manufacturing, the first type of Advanced 3D printing services which appears in our mind usually look like this:

A huge manufacturing laboratory, fulfill by High speed CNCs in one side, 3D printers farm mixed by FFF, DLP and SLA in the other side right next a giant SLM (3D printer of metallic objects). In the backside of this advanced manufacturing laboratory, a pair of robotic arms move object from a conveyor into an automatic painting chamber. In this case, everything is remotely controlled and managed by an artificial intelligence.

Did you see the image? Alright.

In this approach, the technical element is predominant and as a consequence, the fabricated object is expected to be exactly as specified. By chance, the customer will be happy by the end of service delivery. This approach is called *Technical branding service* .

The main problem of this approach is that you have to invest quite much money in order to start having good results. And the worst part is that you cannot even measure the gains and loss of this investment

Why is this a problem?

Because your investment become an amortization amount into your fix costs whereas your customer does not expects to pay for this investment. In the customer's mind, your quality is nothing but your obligation.

In contrast with giant companies, small and medium size businesses have not got millions dollars to invest in advanced manufacturing centers. They have tight budgets and each invested penny need to be paybacked in order not to jeopardize the business itself.

Of course, it is know that big companies such as GE, HP Solid Concept, Stratasys, 3D systems and MakerBot invest millions of dollars in manufacturing labs and technical branding services

And does this approach work?

Yes, it does. Otherwise, they would not have been doing.

The most difficult part of it is the fact that these companies underuse these equipments do not actually know the actual gain obtained from their laboratories.

You, as either a hobbyist or the owner of a small/medium business, cannot afford to spend tons of money in such investment. Thus, you have to avoid the glamour of company like GE.

Ok. You have been imagining : I just want to buy some 3D printers and start selling objects ? It is different from technical branding services, isn't it?

Unfortunately, it is exactly the same. The focus is on technical element (some 3D printers) and when you realize it becomes a snowball that enticed you in a price that is so low you cannot pay your business fixed costs.

Therefore, you need to be more efficient from the point zero on. The Customer driven services is the clever answer for this dilemma .

Please, do not get me wrong. I am not saying that Technical branding service does not work. It works in two scenarios. 1) In the long term, after continuing spending quite much money without payback expectations. 2) In very short term without any competitors.

Nevertheless, None of those cases fits most of entrepreneur, small and medium business. In my opinion, starting a 3D printing business requires that you work hard to ensure that each invested cent become a positive ROI in a sustainable way .

And how can you do so?

In order to better understand how you implement that, you will need to understand what is customer driven services.

"Quality is not what happens when what you do matches your intentions. It is what happens when what you do matches your customers' expectations. "

Guaspari

In this approach, the equipment and technical elements are not as important as the impact that your product/service cause on the customer expectations. Therefore, it is possible to say that the most important part of this approach is to put the customer in the stage, instead of equipments.

Thus, the metric into this approach is directly driven by the return of investments (ROI). In other words, the main idea in this concept is to identify how much the customer is willing to overpay because of the final result of your service. In this case, we will call it **perceived value.**

In the customer driven services , you can measure and directly correlate the **investment** and the **perceived value**.

I will exemplify this using a personal experience. In my first business, I have invested $100.00 up to $100,000.00 and achieved positive ROI.

Do you want to know how I did that?

I created a virtuous investment cycle which works in this way:

Step 1

Identify different fabrication features that excite your customer (Value generation features)

Map thing(features) that people like in commercial unique products, such as texture, tactile, polishing, weight, shape, color, material, et cetera.

At the first time, chose 2 or 3 very simple features to implement. Note that the main goal of the first interactions is not to generate revenue, but understand how it works.

In order to identify these features, you can use friends, clients, and complete strangers. Note that you don't need a marketing research to identify these characteristics.

I started into an appliance store in a shopping mall. Imagine the situation, I pretend to look at some products while I was listening to other customers to talk about product. After that, I (fake) complained about the same features (talked before) and identify the reaction. The reaction people had was to defend their idea(product) and it often exposes the hidden perceived value of customer.

In my case, I began with 3 feature (Metallic cover, shot blasting finish and polish finish). For plenty of people this features are too advanced, for other it is not. Anyway, I was comfortable to develop these features back in the days.

Step 2

Estipulate an investment budget to boost our customer driven service.

As this approach aim to measure profits, revenue and results. The definition of a budget make it realistic and tangible, instead of a non-refundable R&D program.

In my case, the first investment I made was $100 ($33 per feature). Certainly, it might be too much for someone, while it is not a big deal for others. It was an ok value for me at that moment. Moreover, the most important thing you have to thing in this case is to define a budget to start the process.

Therefore, recommend you to define an small budget at the first rounds.

Implement techniques that are compatible with your budget and generates perceived value

In this case, most of features can be implemented by either simple or advanced techniques, which consequently cost different amounts.

At the first interactions, use small budgets because the main goal is to identify perceived features (features that increase the value that the customer is willing to overpay).

Step 4

Measure the impact of features

In this step, you will realize that some features imply on better results than others. In other words, it means that each feature causes a different ROI.

Therefore, It indicates the first direction in the way of increase of perceived value.

Step 5

Redirect the investment/energy from the worst features to the best performed features

Using a metric (perceived value / budget) you can rank the features which performed the best. Thus, cut-off the last place and increase the investment of the first place.

Step 6

Take the part of profit obtained by the previous steps and increase your budget for the Customer Driven Service

In my case, I invested $100 and implemented in products that returned $300. So, I separate $200 of those and add in the my next budget. As a consequence, the next budget sum up $300 to develop and new value features.

Step 7

Repeat previous steps with bigger budget

Back in the days, I repeated all the steps using the new budget ($300) . And believe, it generates profit of $1000, at the moment.

Steps 6 and 7 might look obvious, but you would nearly fall off your chair if you realized how many people who do not follows these steps.

Just to illustrate it to you. One day I met one student of mine. He was part of a course I lecture which is called Make It Stunning. He told me that he had invested more than $200 in some strategies that returned $2000.

Terrific, isn't it? Of course!. So you imagine that he took advantage of this great development and increase his budget. Wrong! He used exactly the same $500 as the budget of the next development.

In my opinion, it was a classic strategic mistake. However, Master customer driven services is not enough to hook 3DP NEC and 3DP NFC. For that, you will need to understand another strategy, **The quality driven service** .

"Quality is more important than quantity. One home run is much better than two doubles. "

Steve Jobs

4

Customer driven services

Vs

Quality driven services

As you saw in the last chapter, Customer driven services was created to measure , improve results and boost impact of value feature on customers.

These features might increase the perceived value in either a incremental or a sensible way. In other words, depending on the type of feature and strategy that you use, the results will be bigger.

In the group of the strategies which imply on sensible perceived value, a couple of step were included into the process in order to boost results. This strategy is called Quality driven services.

Let me explain it better to you.

In the classical approach, you produce objects and deliver to the customer in a direct way. So, the perceived value is included into service in only one step.

On the other hand, the quality driven services affects the 8 dimensions of **perceived quality** (elements in which customers see that worthwhile to spend their money) in **just 3 steps**.

1) Improve perceived features
2) Improve support and productivity
3) Improve services, pre-service and post-service

Ok. Now why is it a game changer ?

Because in the previous case (technical branding), someone order an object to you with close/open specs and what happened look like this:

-There it is the STL. Print in PLA, 0.1 layer height, nozzle 0.2mm ……..

-Here is your object as you asked

-Ok. It is so so. Bye.

Unfortunately, this approach did not bring a good ROI. Then, we had to invest the equipment, optimize machines, upgrade sensors and invest in infrastructure. After working hard and present a new solution for the customer. What happened looks like this:

-There it is the STL. Print in **PEEK, 0.05 layer height**, nozzle **0.1mm** ……..

-Here is your object as you asked

-Ok. It is so so. Bye.

This approach might work well for 3DP EC (3d Printing experienced customer), who already knows what is the purpose the object and the 3D printing parameters that are enough for them. This sort of customer would usually rather you to take this approach. In addition, they have their own 3D printers in quite many cases.

Nevertheless, this approach does not work for 3DP NEC (3D printing non experienced customers) and 3DP NFC (3D printing non familiar customers). The worst part is that these customers usually hate technical branding approach.

Applying quality driven services, you build a strong relationship with potential and current customers. In this case, you over deliver to your customer without charging a penny.

Another interesting point in all of this is the fact that Quality driven services increase perceived quality in only 3 steps, in addition to building another very important asset: direct and indirect partners.

Well, why does this sort of approach work so well?

For better understanding , let's remember the 3 types of customers

- 3D printing Experienced Customers – 3DP EC
- 3D printing Non Experienced customers – 3DP NEC
- 3D printing non familiar Customers - 3DP NFC

Do you also remember that each 3DP EC correspond to 20 3DP NEC and 50 3DP NFC?

Thus, Quality driven services target exactly the portion of 3DP NEC and 3DP NFC.

It happens because these customers do not want to fabricate the object straight way. They want to you to help them to develop and launch their IDEAS.

They do not want to spend money on work that you put on their ideas at the beginning. But, after you help them to maturate the idea ? or After you help them to create a stunning feature, such as a tactile texture, which will transform their idea in the product? Or even Or even after they trust in your competence, productivity skills, organization and creativity?

Only after you are building this relationship with the customer, you will understand their need and expectation. And when they decide to fabricate, they will not make an 3D printed object (cheap, brittle and not worthy). They will fabricate a Product (more expensive, robust and relevantly more valuable) which occurs to be produced in 3D printing (at the beginning).

I will give you one example of that for you to understand better.

Imagine the picture:

 A store owner hire a genius painter (a genius in the level of DaVinci or Raphael) to help him to paint his store.

Before starting, the man give the painter the blueprint marking exactly what is the color and paint which the painter should use.

The change of the man being satisfied is statistically minimal. Because he had hired a genius which delivered a work which he could do himself.

A better way to achieve success in a case like this would be:

The genius painter help the shop owner to understand architectural concepts

After identifying the expectations and needs of the man, the design is elaborated altogether. As consequence, the product plan is created and the benefits of the concept is significantly better than the first raw and abstract idea from the beginning.

Therefore, the store owner can hire the painter to make the design without hesitation. The painter produce a master piece which will make the owner happy.

Does it make sense for you?

If it makes sense, hold on, because now we will learn everything about perceived quality.

NOTEPAD

2

How to implement the approach of Quality driven services

(For new or current businesses)

The most precious asset of yours is your client

It is interesting to note that both customer driven services and quality driven services put the customer as an important piece of a complex beautiful machine.

Collecting data from customers are excellent tool to guide your next steps in the perceived quality and decide how to prioritize your investments.

Uncontrolled Clients

This type of client is very interesting and as any other client, they invest into your company. They believe in your work and sometimes have periodic orders.

The problem in this type of client is the fact that they don't share information and they don't want you to understand their needs and help them to build products.

They have the relationship control over you. In addition, they can even move to other competitor without hesitation.

In other words, the risk that you will face in front of this client is high. So, I recommend that you keep this clients running on while you develop new relationship with another client that you have control.

Since the first industrial revolutions, one thing is certain. Collaborative work and collaborative design had been ensuring manufactures to grow as satellite of their clients.

For example, an automotive industries are nothing without engine factories or wheels factories.

Therefore, it is possible to say that a controlled client is the one who **shares project and design responsibility**.

You understood what I am going to arrive, didn't you?

In the moment that you start helping you clients, you become their partner. Therefore, Quality driven services start to become reality.

In order to achieve challenging targets, powerful tools are required. I will present to you the 3 tools that will allow you to understand your controlled client.

- 8 Dimensions
- QFD – Quality Function Deployment
- Blind Experience

In general way, the perceived quality has 8 dimensions,

- Dimension 1: Performance
- Dimension 2: Features
- Dimension 3: Reliability
- Dimension 4: Conformance
- Dimension 5: Durability
- Dimension 6: Serviceability
- Dimension 7: Aesthetic
- Dimension 8: Perception

As we learned, each customer sees the worlds from one different window, but all of them instinctively take attention to all dimensions, even though this dimension is not important to them.

Performance indicates that the product works in the way it was supposed to work. It is important to note that this is not an obvious matter. You have to considers that in spite of errors of your customer design, success of your service will depend on how well the product works or how well prototyped parts assembly each other.

In this case, try-out tests make all the sense before a final deliver.

Features are things that generates value to the product. This dimension is the one you should exploit the most.

Each design has special characteristics to become unique. Each designer has his/her either.

As features are small parts of design, you can improve it progressively and incrementally, such as we presented in the strategy of Customer driven services.

Reliability is the things and elements which make your customer to trust and rely on your services. This is a branding characteristics that is fundamentally defined during the first services or by indication.

In this case, deliver features that ensure that your product is reliable make plenty of difference.

I will give you one example. Once, I implement the policy of delivering a dimensional/assembly/tensile testing report in the delivery product/parts. In this case, the customer which I aimed to increase reliability has the dimensional/mechanics characteristics as his top priorities. Instantaneously after implement this policy, they changed the behavior and start openly discuss design and projects with our team.

Gain the trust of whom you want to be partner.

Conformance is the characteristic that 3DP ECs like the most. It is related to the precision and how alike the physical object is from the design and 3D model.

In this case, it is important to ensure a suitable conformance. But pursuit of great conformance usually implies on equipment, training and labor investment.

In this matter, you have to realize that most of **3DP NEC and 3DP NFC don't care about conformance, but performance.**

In addition, controlled clients share responsibility with you, giving you opportunity to adjust the design in order to absorb tolerances and fabrication errors.

Durability is a factor that is highly enhanced to cost and price. This dimension describes how long or how much the product will take.

In this case, nothing is eternal, but disposable thing sounds cheap in customer mind.

Therefore, you will have to find the balance between disposable and eternal levels.

What I recommend in here is to start with disposable materials, configurations and poor finishes in the internal try-outs. And use a good materials and practices in the external (showing to client).

Never let your client keep a poor try-out (prototype) even it is at the beginning of your development.

Further information can be found in material strength books (very technical) or in the book Make it stunning.

Aesthetic is the dimension that make the eyes of everyone shine. In this dimension, the first thing that you have been imagine is a polish surface.

Of course, some one that is also familiar with 3D printing techniques will also say. Use acetone or methyl Ethyl Ketone (MEK) in an ABS object.

By 15 years of experience in 3D printing segment and being a winner of the international award of best 3D printing research, I am affray to say that it is a rookie mistake.

I will explain it to you so that you will understand. I am not against solvent vapor attack (such as acetone vapor). I am saying that most of customers don't like it. They are looking for innovation and combination of senses. And because there are infinite types of finishing, materials, textures, techniques and tactile elements, the both your mind and your customer mind can freely flight and create an unique texture feature that will make your customer product special.

Serviceability is the dimension which indicates how easy the product is to be fixed or repaired.

It is interesting to note that in spite of the importance o this dimension. I can say by experience that most of customers don't care about this quality dimension in the stages of the product development.

Nevertheless, a suitable design of multiple parts product, in addition to joints, fasteners, snap fits make the difference between amateur and professional designs.

Quite many recommendations about this matter can be learned in books of machine design. It will help you to have ideas about attachment systems.

Perception this dimension is the most important dimension in most of cases. This dimension can also be called Wow dimension.

Now is the moment that you use all the powerful knowledge about your customer and put into the product. Fabricating personalized things for your customers make them feel important and special.

Additionally, this dimension is widely cover by the co-responsibility approach, where you help your client to design the product.

That is one of the reasons why a solid relationship with you customer is so important.

> "If your customers like you (your services), thing that you make will easily please them.
>
> If your customer do not like you(your services), any work that you make will not please them."

Ok, now that you learned what are the 8 dimensions of quality, how does each dimension can be used in your favor?

One of the most common methods to put in practice is 8 dimensions map. In this map, it is possible to evaluate the effect of feature on your customers.

This method is a powerful tool to identify how good is the impact of a developed feature on your customer, and you can do it in 3 steps.

Step 1

Therefore, you evaluate how much your customer value each one of 8 dimensions in a scale from 1 to 5.

Step 2

Analyze value the feature or product in accordance with each one of 8 dimensions in a scale from 1 to 5.

Step 3

Find the impact coefficient by multiplying the values of features and customers.

Values higher than 9 indicate that the feature will cause good impression on your customers in that dimension. Values lower than 9 will probably be ignored.

For example, once I implemented a new rubbery texturized finishing as part of the brand new development of a customer design. I evaluated a 3DP NEC who aimed to proof a design concept.

In this case, the strongest characteristics of the feature that I was implementing were aesthetic and feature.

After mapping both the customer and the feature, I was able to predict that the customer perception about the product would be Aesthetic Feature and Perception.

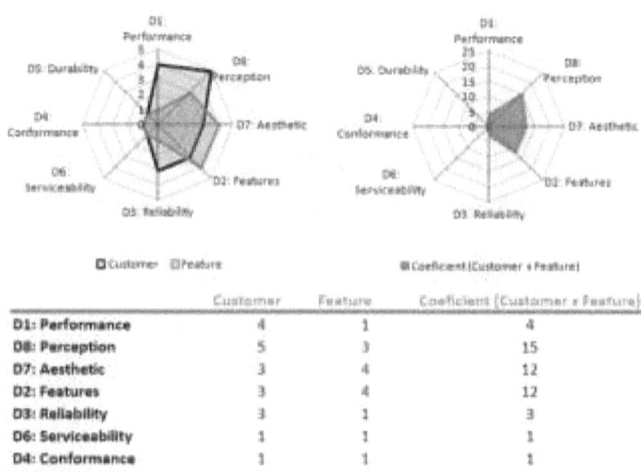

	Customer	Feature	Coefficient (Customer x Feature)
D1: Performance	4	1	4
D8: Perception	5	3	15
D7: Aesthetic	3	4	12
D2: Features	3	4	12
D3: Reliability	3	1	3
D6: Serviceability	1	1	1
D4: Conformance	1	1	1
D5: Durability	1	1	1

Thus, what does it mean?

Delivering an stunning object in terms of aesthetic and tactility which make his product different from his past experiences would achieve 12 times more impact than fabricates a perfect object in terms of dimensions and strength.

Well, you can easily see that 8 dimensions might imply on quite many qualitative characteristics , characteristics with cannot be counted or measured in numbers.

For example, you evaluate a customer that indicates that he want that his product/object is beautifully shinny. Certainly you identified that he considers aesthetic and feature as the most important dimensions of quality.

But, how shinny must your object be?

If you make it right once, how can you repeat the result? How can you measure when you can stop?

In order to quantify these uncountable (qualitative) characteristics, I will show you the tool that revolutionized automotive industry: Quality Function Deployment (QFD).

Quality Function Deployment (QFD)

This tool is one of the most useful approaches to connect marketing analyses and technical specification. It correlates the needs and expectations of customers and the technical characteristics of design.

In other words, this method indicates that, **beautifully shinny** can be **measured by roughness and reflectance**, while it can be **achieved by Surface treatments** (such as chemical attack, thermal attack, coating, glazing, among others).

Technical Similarity

Design Features

Customer Expectations

Correlation between expectations and Features

Benchmarking

Results and Priority Ranking

In a simple way, the QFD can be described in 3 main parts:

- Customer expectations
- Design features
- Correlation between expectations and Features

Illustrating the application of this table, I will show you the analysis of a simple Mug which a client of mine asked for my help to develop.

I this case, the customer expected that Mug should be:

- big
- Fit in the car support
- Look like himself
- Don't drop coffee into car because of motion
- Warm hand
- Don't burn hands
- Keep coffee warm

In contrast, what did I have into my toolbox to develop this product?

I could define:

- Internal mug volume (Oz or ml)
- Size (cubic feet or mm^3)
- Color
- Material
- Shape
- Coating and painting
- Insulation
- Lid

Perfect, knowing the importance that each expectation of my client, I could correlate the design feature and customer expectations .

I found that the 3 most important thing in the design from the technical point of view were:

- Shape
- Type of material
- Insulation

In contrast, this exactly design characteristics made my customer to see that the mug:

- Don't burn hands
- Keep coffee warm
- Look like myself

Strong Relationship - 9
Medium Relationship - 3
Weak Relationship - 1
no relationship -

Item	Weight / importance low - 1 / medium - 3 / high - 9	Customer Expectations \ Design features	Internal Volume (Oz or ml)	Size (cubic feet / mm³)	Color	Type of Materia	Shape	Handling /Painting	Insulation	Lid	Weight Score	Relative Weight Score
		Business Priority low - 1 / medium - 3 / high - 9	1	1	1	1	1	1	1	1		
1	3	Big (fill a lot of cofee)	9	9			3				63	6%
2	3	fit in car support	3	9			9				63	6%
3	9	look like myself			9		9	9			243	25%
4	9	don't drop coffee into car (because of motion)					9			9	162	16%
5	1	warm hand				9	3		9		21	2%
6	9	don't burn hand				9	3		9		189	19%
7	9	keep coffee warm				9			9	9	243	25%
		Weight / importance	36	54	81	171	228	81	171	162	984	100%
		Max Weight in the column	9	9	9	9	9	9	9	9	984	100%
		Relative Weight	4%	5%	8%	17%	23%	8%	17%	16%		
		Ranking	8	7	5	2	1	5	2	4		

It is almost evident that using this tool to develop a mug is like killing a fly using a cannon. Nevertheless, this example evidences how useful and powerful this tool can be to improve each aspect of your product, design or business.

Ok Marlon. This looks like too theoretical to me. How can I try another hands-on approach?

Certainly, the next method that I will show you is the tool that you should use to collect and evaluate data, mainly for the last methods.

This method is an amazing empirical tool which helps you to improve the quality of your products/services in quite few interactions.

Have you ever heard about Placebo effect?

In medical and pharmacological developments, they test the effects of drugs in humans. Nevertheless, people are complex and difficult to measure. Because of that, part of drugs pills of trial are placebos . They look like the actual pills, but they are not (can be sugar, flour, et cetera). The only ones who know which pills were placebo are the researchers.

In the blind experience method, the placebo effect is the driver motor of method.

This method consist in 4 steps:

Step 1

Prepare 3 products/ objects where each one has a different features to be evaluated.

Do you Remember the customer driven service? That is the time that you can apply a lot.

It is important to note that the only thing different among the objects must be the feature.

Step 2

Prepare 2 objects with **NO FEATURE.**

Exposes only 3 of products/object in a small group of clients (3 to 5 people)

Try hard to select all from the same type of customer. Give preference to 3DP NFC.

Check which one they like the most.

Replace one of products and repeat step 3 again until of parts have been compared.

It might finish in 5 to 7 interactions.

By the end, you will have the score of each feature that you are evaluating so that you can rank and improve your techniques.

2

How to improve productivity and support

Certainly, costs are extremely important in a business, project or design. In addition, this matter is the difference between life or death from the economic point of view.

As you learned before, perceived features and perceived quality are the main elements that justify your customer to spend more for your services or products.

For that reason, the you can increase your offered price without jeopardizing your customer.

In spite of that, it is important not to be trapped by productivity. How is that possible ?

Imagine this situation:

You start producing an outstanding product that pleases your customer.

The price of your product is $100 and you sell 10 products per month. Imagine that you sell only this product.

Therefore, the business incoming is $1000/month.

However, you spend $900/month to fabricate the products in accordance with the perceived quality.

- Office rent - $500/month;
- Electricity - $100/month
- Material - $200/month
- Coating /sanding papers/etc - $100/month
- Equipment amortization - $100/month

By the end, you will have a profit equal to $0/month.

That is a hard situation, isn't it?

That is the most common trap that helped 80% (around) of 3D printing services since 2013 to close. But, don't worry. I will help you to escape from this threat.

Alright, the 2 most important tools to increase your profit in addition to creating a remarkable impression for your customer (increase the perception dimension) are:

- Lean Techniques
- Advanced manufacturing techniques

Lean techniques

These techniques born inside Toyota and help them to become the biggest automotive company in the globe. As a consequence of the big success of those techniques, almost all automotive companies adopted lean techniques, nowadays.

The unbelievable part of this story is that the main core of lean techniques can be easily implemented by only 3 principles:

- 5 s
- Standardization
- Value Flow

Apart from the mislead mind image created by popular culture. 5s is not cleaning.

So, what is 5s and how to use 5s to improve my productivity.

5s is a method which imply on the mindset change of an entire company. It is divided in 5 steps:

- Sort (Seiri)
- Set in order or organize (Seiton)
- Shine (Seiso)
- Standardize (Seiketsu)
- Sustain or self-discipline (Shitsuke)

1st step (Sort)

In this step, the main goals are focus in the reduction of distraction and increase the usefulness of tools and equipment that generate value.

I will give you one example of how Sorting help you to improve your productivity. Imagine that the tasks and tools that you need to fabricate an object are numbers.

How much time do you need to count how many number there are in the image before sorting.

Use the timer to help you.

Does it take to long?

Alright, now count how many number there are in the image after sorting.

Use the timer to help you again.

How fast did you finish the task?

In average, sorting reduce the task timing in 10 to 50 times.

Before Sort

After Sort

In order to implement this step, I will teach you a fast technique to make it happen.

1) Put all the thing that you use in just one place (it will become a messy at the beginning but I will worthwhile)

2) Discard all items that you don't use for more than 6 month. Separate the items that you don't use in 2 month and label it with a red tag. Put all red tagged items in a corner.
3) Classify all useful items in accordance with the shape, application and how often you use them. Label them with green, yellow and grey tags.
Green tags are items that you constantly use (more than once a day)
Yellow tags are items that you use not so often (more than once a week)
Grey tags are items that you don't use often (more than once a month)
4) Now that you have your items sorted by frequency, shape and application, put each group in one box
5) Green tags will be next to you while yellow tags will be on the shelve in a accessible position. Grey tags will be on shelve in a non-accessible position.

2nd step (Set in order)

Set in order is centered in the fact that that the workflow should be smooth and fast in order to reduce time wasting and increase productivity.

The basic stages to achieve an good organization are:

1) Arrange all the items that you need in order to be easily fast to find and use.
2) Define what is the shortest location for each item or group of items that you use in the process.
3) Define fix location for each item and group. Each item need to have its own place.

Using a simple 3D printing shop as an example, I will show you how Set in order make the fabrication agile and faster.

Imagine that you are producing 3D printing objects in 3 working stations where one of them is a painting station.

The basic workflow for each part usually follows:

1) 3d printers
2) Finishing station
3) Painting station
4) Drying chamber
5) Inspection and shipping

Without setting in order, the workflow is confuse, people crash each other and it looks like there is no space to do what is needed to do.
In this environment, people are stressed, make tons of mistakes and rework things all the time. The time spend in each task is insanely long and production cost is high as a consequence.
On the other hand, setting in order the same space and working station will cause a miracle. The workflow become clear, short and safety. Environment become clean and people produce faster, happier and healthier.
By experience, setting in order reduces rework and wastes in 80% in addition to reduce production time in almost 40%.

Magically, the space become larger and less claustrophobic so that people can think and work better and faster.

3rd step (Shine)

In this step, it is not just about clean the environment where you work.

Ok, but it is not about cleaning, what shine is about?

It is about keeping the environment constantly organized, clean and sorted.

In other words, make it shine means that **you have to continually maintain sorting and Setting in Order**.

Of course this also indicates that you have to clean it in the daily basis, otherwise you won't be able to maintain everything organized and sorted.

Another point that makes Shine so important is the fact that the dirty produced by 3D printing, finishing and painting jeopardize the final state of objects that you fabricate.

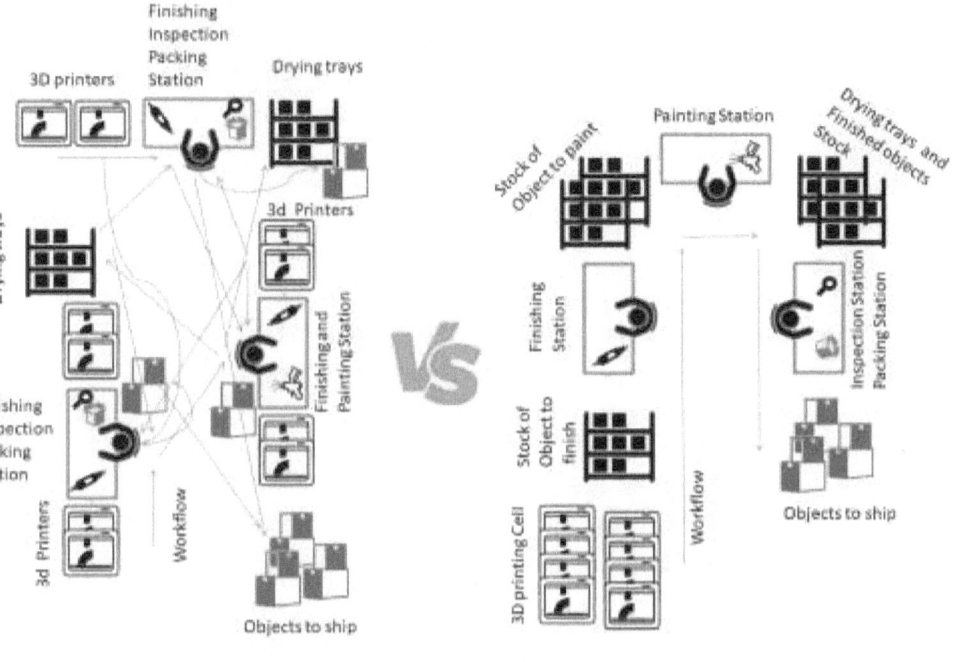

Before SET IN ORDER

After SET IN ORDER

Imagine that you want to eat an delicious creamy cheesecake which was just made by someone which you considers that cook very well. My situation, I can say that this person is my grandmother.

You smell this insanely delicious cheesecake and you urgently grab the first fork that you see in front of you and take a piece of this creamy cake. Unfortunately, You grab a dirty fork that was used to prepare onions and gorgonzola cheese.

Instantaneously you lost the desire for this cake, don't you?

That is likely what happen in making stunning 3D printing objects. The dirty spoils the object and the desire of making this object outstanding.

Now imagine the same situation described before. But instead of grabbing any fork, you special cooker (my grandmother in my case) bring a slice of the cake with a fork and cinnamon on the top. Oh my God, I am almost calling my grandmother now just because the image that I am trying describe to you.

Alright, to make shine happen, you will have to:

- Keep everything in its own place
- Clean dirty frequently
- Check the maintenance of equipments during cleaning.

I recommend you to use a fantastic technique which quite many genius entrepreneur use. The Pomodoro technique is a method where you continuously work with no disturbance during a long period of time (often 25min) and stop during a short period of time (often 5min). Each cycle is a pomodoro.

Thus, each 3 pomodoro, you take break of 15min.

Easy, isn't it?

I recommend that you follow:

> **2 pomodoro working**
> **1 pomodoro cleaning, maintaining and planning next tasks**

4th step (Standardize)

Standardize is one of the biggest tools in the industry. It allows you to achieve repeatability whereas the main goal of this step is to create and document procedures of the last 3 steps.

In other words, this step aims to create signs and step-by-step procedures that anyone can follow, in spite of being a beginner.

Therefore, I will show you the basic guidance to implement Standardize in your daily routine:

- **Define a organizing box:** select just one type of organizing box and use it for everything
- **Label everything** : boxes, working stations, tools, locations where things are supposed to be
- **Organize shine tasks:** define areas where everyone is responsible for. In addition, define a

schedule of cleaning where everyone works together.

- **Use photos:** use photos to describe procedures and Remember the progress from one week to other.
- **Score and improve:** make everyone to evaluate the environment and indicate new ideas to improve. Try to make it in 15 minutes meeting each 7 to 15 days.

5th step (Sustain)

In this step, the main goal is to ensure that the 5S program keep working, improving the environment and increasing productivity.

Plenty of options are used in this step. Nevertheless, I will present to you one technique that is easy and I am sure that works.

The 5s meeting with kaizen board (board of improvements) is one technique that is used to record improvement ideas that we briefly think, but because of daily working we forget.

In this board, anyone should give one idea in each 2 weeks to present in 5s meeting. The idea is just to write a brief improvement that can be implemented.

Remember that 5s meeting is very fast (15 -30 min) so that I recommend that it follows the sequence:

- Describe last status Vs the current Status
- Define the current Score

- Present notes from kaizen board.
- Vote for the top 1 or top 2 kaizen notes
  - The selected kaizen should be implemented in the next weeks
- Define new metrics for the follow evaluation
- Celebrate achievements

What should be included in the kaizen cards?

In general it can be anything, but the most efficient subjects to be treated in this cards are based on the 8 wastes method.

The 8 wastes

In this method, the lack of productivity is described in accordance with 8 type of wastes:

1) Defects
2) Inventory
3) Motion
4) Waiting
5) Extra-processing
6) Transport
7) Overproduction
8) Non-used Talents (skills)

It can be noted that all these wastes will affect the cost of fabrication, quality and delivering time.

Therefore, kaizen (improvement) notes should aim to solve just one problem at time. And after sometime, all the matters would be addressed and your product(service) will reduce costs and increase perceived value.

I will give you one example from the kaizen board of one of my first businesses.

We were producing 3D printing objects in a very high volume and from time to time, one phenomenon happened. Periodically, there were moments that our team was overloaded, and moments that they were absolutely idle.

How Is that possible? The workflow considered that each one finished its task as soon as possible. No matter what.

The biggest problem on that was that the fastest way was not the most suitable way, in addition not to being the most effective way.

Therefore the kaizen cards was included where we implemented the production order which counted estimated time for each task of the entire process.

Therefore, it was possible to organize activities which were similar and create a daily working pipeline.

For implementing that, we had to

- measure the average time of each basic task of our daily basis
- Create a production order which describes all the required step to produce an object/product.
- Calculate Takt time (what is the longest that will still please your customer)

- - Takt time = Available hours / customer demand
- Create the schedule of task for the day based on Takt time

In this case , the balance between tasks were defined by the technique called **OBD (Operator Balance Diagram)** .

In this method, the time that each working station or operator (person) spends is presented in order to see who is overloaded and who is underloaded. Therefore, it is possible to reallocate someone to help other position and share the burden.

For example, imagine the following situation:

You receive order for 70 medium products/week where each part consume:

10 hours in 3D printing

2 hours in preparation and sanding

2 hour in tumbler finishing

10 min spraying primer

30 min drying primer

10 min in painting 1st hand

20 min in drying 1st hand

10 min painting 2nd hand

1 hour drying painting

Imagine that you have 4 3D printer working full capacity for you besides counting on 3 people in your team.

Thus, your takt time is going to be 1.7h/product. As a consequence, you can start analyzing by your basic configuration, and then balance (share tasks).

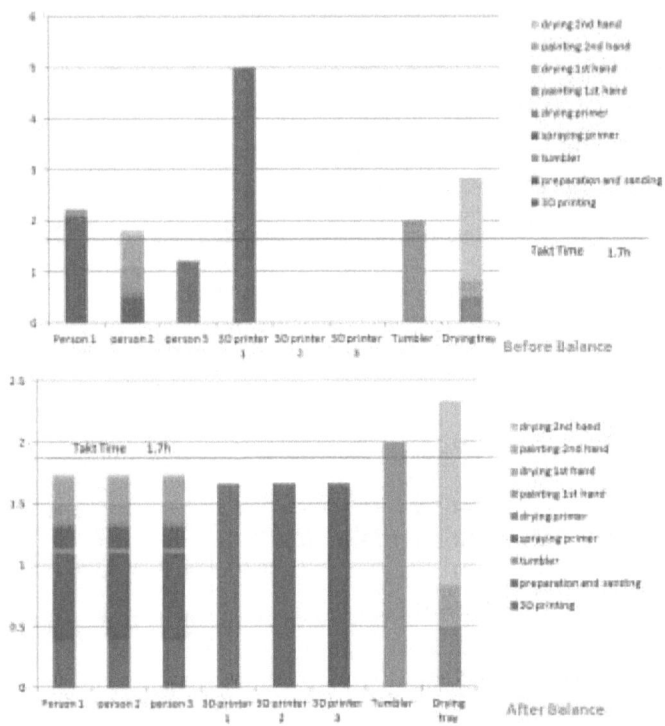

In this example, 3 3D printers is enough to attend Takt time, and after balance, the same 3 people were found to be enough either.

Advanced finishing Techniques (the Holy Grail)

The third technique is what I nicknamed "the Holy Grail" of 3D printing manufacturing. This is a strategy which is by far the one which most bring results for me and my customers.

Advanced finishing techniques are a very complex matter and, by experience, caused the most amazing results that I could ever get.

In order to consistently achieve outstanding results applying advanced finishing techniques, I created a guideline which compile the most advanced techniques that will lead you to make every single product of yours stunning.

Now I have a very good and one bad news for you.

The bad news is the fact that advanced finishing techniques is more well elaborated strategy and is not possible to explain in only one book chapter.

The good news is that every year I launch free workshops to show how some of the techniques work.

If you want to make your free inscription, I will send you the year schedule and you can chose which date is better to you:

www.makeitstunning.com/workshop

Of course, the most interesting thing to do is to start applying the knowledge you learned in this book because it will bring you new perspectives to use in the workshop.

In addition, you can also find the handbook of the advanced finishing techniques for 3D printing technologies in the book:

Make It Stunning :A concise guide to finish 3D printing objects

3

The black box secret

In this chapter, I will show you something that you probably was about to discover. How the quality driven strategy works.

In this black box, you can boost results applying some pieces of advices which I will show you.

First part of this is to create the relationship with you customer. In this case, you will have to work hard to create new and innovative solutions that help them to create products.

Implementing the techniques which I presented you in this book about quality driven services will drain quite much energy. But it worthwhile. Always remember that it is not a single shot technique, it is a progressive process which requires that you are resilient to implement.

The second part of these process is to understand exactly what are the customer needs and expectations. In this case, you have to show options in order to collect information. Advanced quality techniques were introduced to you, so start collecting data and understanding how finishing features , design features and service features will impact your customers.

The last part is to exceeds your customer motivations. In this case, I always indicate that after deliver a very good product, it becomes outstanding when you overdeliver or when you give a gift feature which will trigger the essence of your customer.

How to create a successful overdeliver

As we saw in this book, each customer sees characteristics that drilled the foundation nature of theirs.

Therefore, the overdelivery need to be exactly in the "g-point " of its perceived quality.

For example, if you identified that the customer has the strongest tendency pointing to aesthetic dimension of quality, give him a free product with an alternative finishing or shape.

If you try give him discount, it will not bring you any benefit.

Therefore, don't work with low profit margins, otherwise, overdelivery won't be able to do done. I always recommend my clients to consider 20% of profit to overdelivery solutions.

Be a hitchhiker

It is obvious that you need to be creative to create new products for others. So, create a vibe of developing products and create your own products.

They will help you to create portfolio and create new technique by your own. I will help you to test and financially support your incremental investments .

For example, once I developed some unique hanger for a customer. But during the development process I faced several ideas that were not Hangers but could be developed by the same techniques which I used to produce the hanger.

Great, isn't it?

So, I created a board of ideas which anyone of my team could include new cards. Once a week, I discuss with the team and they can produce the idea, sells and get 90% of profits of this product.

In addition to being a motivation tool, it creates an innovative environment where everyone wanted to create their products and sells online.

NOTEPAD

3

Do you want to flight higher?

Learn how to build a low cost High Tech Lab

1

How to select which equipment will bring you value ?

As most of entrepreneurs, each cent that you invest need to transform in profit. Therefore, it is normal to start from the basic equipments and upgrade your fabrication site on the road.

But what is the best strategy to invest?

It depends on the perceived quality of your customers.

I will give you one example of one of my customers which become a friend after some time.

I met him in a conference when I was presenting new techniques to include physical solutions into digital world.

At that time, I lecture a course about how to design, assembly and operate 3D printer.

During that conference he insisted that I lecture him a whole different course. He knew the basic of 3D printers but not how to create 3D printing shops or fablabs.

After he convinced me to create this course, We started to develop a long advisory program. In this program, We started with the basics, 3D printing an after mastering 3D printers, We would move to perceived features and perceived quality by the end.

As a consequence, the equipment of him would be upgraded in accordance with the money income and customer understanding (information).

It was an amazing journey where he started with 1 3D printer and finished with 10 3D printers, 1 tumbler, 1 finishing station, 2 casting station and 1 painting station.

Remarkably, everything started with the need for automotive spare parts. The most important features for their customers were based on aesthetic and conformance dimensions. Therefore, he invested in techniques which increase aesthetic features and do not jeopardize dimensions and strength. Back in the days the features were:

- Polish mirrored
- Dull and matte surfaces
- Diamond Texture
- Random Texture
- Dimension precision
- Strength

Thus, the dimension and precision were the first features that he solved. Using carbon fiber filaments with homologated supplier implied on strength and almost no distortion even in open chamber 3D printers.

After that, he applied the strategy customer driven services and discovered the strategy to buy the equipment he would achieve great results.

For dull and matte surface, he acquired:

      Hand grill grinder

      Finishing station

For polish mirrored surface, he invested in;

      Painting station

      Airbrush and compressor

      Turntable

      Hooks

On the other hand, the diamond textures were obtained by knurling and embossing tools which he fabricated by himself.

After some time, he upgraded his finishing station with tumblers and shot blasting guns in order to reduce finishing time in half , besides increasing perceived quality.

As this is a very good case study, I recommend you to follow the same steps, considering your perceived features are the same of the ones he had found.

How to decide what to make or buy?

Make or buy is an incredible difficult decision in the business. Bringing this dilemma to equipment investments, the first thing that we imagine is BUY.

Unfortunately, several times you won't have money to buy the equipment that you want.

In this case, there are several open-source projects which will help you to take your step further.

Moreover, remember that your business or your hobby is making things. Therefore, making will just imply on time, fix costs and raw material.

For example, if you discover that a laser engraving machine will bring a huge value for you product. You will have 2 options:

Buy it by $500.00

Make it by: $100.00 + 10 working hours.

If you have an intern to do so, 10 working hours of him/her will be almost nothing in comparison with the $500.00 + training and shipping.

On the other hand, if you have orders that will bring you a profit higher than $400, buying is the best solution.

In order to find opensource project to assembly a laboratory, I always recommend for beginners the book:

Open-Source Lab: How to Build Your Own Hardware and Reduce Research Costs

Otherwise, other advanced opensource projects which is related to advanced finishing techniques is presented in the book:

<u>Make It Stunning :A concise guide to finish 3D printing objects</u>

In spite of buying or making, the most important thing is to invest in equipments that will bring value to your product.

3

Let's hand-on

I know that start or change the company mindset is not an easy task. Each step is difficult, but following this step-by-step the whole mission is going to be easier, isn't it?

Remember that I said before:

"In life, some tasks are easy, even though they are difficult. Thus, here we are.

The biggest shot in this case is that I am giving you the tool which you need not to struggle and suffer as I did in the past. Therefore, you can accelerate your progress and save energy to apply in your business and projects.

$\rightarrow$ A step further

Did you arrive here? I have an special gift for you!

www.makeitstunning.com/Gift

"Quality is to give the customers what they want. "

Sam Walton

NOTEPAD

www.ingramcontent.com/pod-product-compliance
Lightning Source LLC
Chambersburg PA
CBHW030716220526
45463CB00005B/2064